WIND POWER:
WINNING CONCEPTS

TIPS FOR INVESTORS

DR. P. RAJAN

To my wife who inspired, supported and encouraged me with love
In my grief
In my depression
In my anger

CONTENTS

PREFACE

This book, written in a simple and eloquent style, deals with the complete description of a windmill project. Starting with the site identification, it leads us through the choice of a wind turbine, its parts and functioning and the actual setting up process, the related power grid integration, the cost and the probable benefits to the aspiring investors.

A Win – Win cash cow investment plan for investors thinking high with tax holidays and other paybacks, with the added benefit of environmental conservation. Wind energy business potential and prospective business models for the benefit of entrepreneurs are detailed. Further the probable environmental issues that may cause time delay and cost overrun of wind farm projects are dealt with.

A touch of ancient history of wind revolution, its transformation to the present modern form, global development and latest trends add beauty to the book.

◆ ◆ ◆

A MIRACLE OF NATURE

From time immemorial, you pass by
You were Omni present, yet unseen.
Through the occasional storms you try
To show your mighty form so unforeseen.

We seek to conquer your spirit and bulk
We try to rein in your sweeping form.
Like a parent amused at her child's prank
You let us play our mischief in your arm.

The strength so staggering in thy flight,
The tenderness and love in thy nurture,
The powerful force, massive yet so light,
You, oh wind, we seek to capture .

Your soul and energy to brighten,
This world and light up our lives.

Dr Minthu Rajan

ACKNOWLEDGEMENTS

I am indebted to my colleagues and senior consultants for their constant support and guidance without which I would not have been able to complete this work. I also express my sincere thanks to the experts on wind energy, Mr. Illinge and Mr. Hanse. This book would not have been made possible without their inputs about wind mill technology, who have shared their knowledge with me via phone conversations and E mails.   I acknowledge the valuable assistance given by following persons/ Institutions / organizations /manufacturing units for their kind co-operation in supplying various technical data and guidance. American wind energy association, Department of Renewable energy India, C-WET Madras, Suzlon energy Ltd, MDI library, Gurgaon and teaching faculty at IIM Bangalore and TERI (The Energy Research Institute, New Delhi). My special thanks to Amazon.com for marketing this book.

FROM THE AUTHOR'S DESK

T his book on Wind energy was born out of my passion towards nonconventional sources of energy. Basically I feel I was in constant search of sources of alternate energy, given the circumstances of my childhood. Born and brought up in a remote village of "God's own country" named Kerala in India, I grew up studying with the help of kerosene lamps. Electricity had reached only the towns and cities. We had to depend on firewood for cooking. Electricity remained a distant dream in my village.

During my under graduation in Engineering, an article in college library caught my attention. It was a data on rural electrification, about United States of America, setting up wind mills to electri rural households in California. This idea mesmerized me, as the thought struck me that wind was a natural gift in my village and still we remained in darkness.

We used wind for drying our cloths, drying our grains and for pumping water. Soon after graduation I took up the job of electrical engineer in government sector and was posted in my own hometown. By then the revolution of using non-conventional energy sources for electrification of remote villages was slowly taking birth.

My desire for electrification of my village drove me to take a doctorate on the subject and later I did my post-doctoral research on this field. My knowledge was put to good use during my career. Now, this day, we have installed wind turbines sufficient to meet the energy demands of remote villages in the beautiful hills of Attapadi, a part of the western Ghats and many other neighboring villages.

While working on wind farm projects of our locality, we faced a lot of challenges. Objections arose from several quarters including the natives and nature lovers. Many of the apprehensions were out of the ignorance about the wind turbine technology and its functioning. Fears about shortage of rains, decrease in the natural flora and fauna of the place, seizure or loss of the land, loss of their livelihood; all had to be discussed and sorted out.

This book seeks to salvage all the fears and apprehensions, to provide some basic information about the wind farm projects. I have incorporated my field experience of wind farm construction into this, expecting it to be an inspiration and guidance to all young aspiring wind farm engineers and probable entrepreneurs in this field to face the problems that they might encounter at their work site.

ENERGY AND ENVIROMENT

E nergy is the driving force behind the industrial growth and economic development of any country and developing countries in particular. A reliable and affordable energy system is a must for sustainable GDP growth of a country and for the better quality life of its citizen.

Right from the beginning of human civilization energy been used by human being all over the world to meet the requirement of daily life and basic needs. But in developed countries natural resources are being utilized more for comfort and luxury. The burning of natural fossil fuels for energy is causing air pollution and other environmental problems. The reserve of conventional resources is finite and it is utilized unscrupulously by mankind, saving nothing for the future generation.

In the present scenario, the thought of setting up a nuclear power plant or super thermal plant in a densely populated country is quite impractical. Dependence on Hydel power plants will invite environment challenges like huge dams, threat to the wildlife and submergence of forest land. The extent of forest land is decreasing day by day, thereby reducing supply of oxygen. Moreover, the increasing energy demands of world cannot be met solely by conventional sources alone as its reserve is limited. The energy derived out of fossil fuel may not last long and it is high time that we should think of alternate source of renewable and environment friendly energy.

In this context harnessing of renewable energy such as wind or solar could play a significant role in the energy mix of a region. Wind energy has been used for centuries to grind grain and pump water in rural areas. Wind energy is renewable and environmentally benign.

Compared to other conventional energy sources, wind energy has the added advantage of being investment friendly. Wind energy can diversify the economies of rural communities adding to the tax base and providing new types of income for investors. Wind turbines can add a new source of property value addition in rural areas that have a hard time attracting new investors. The wind power policy allows accelerated depreciation of the wind power plants.

The investment on wind projects with cost out of the

profits accrued from other businesses of a company, helps to cut down overall tax liability of the company. Businessmen are using this window for reducing tax liability by installing cheaper wind mills of out dated technology based on supplier's claims without any market research. This method of wind projects financing and implementation is leading to inefficient harnessing of the available wind potential. This type of investment model has become very common among entrepreneurs all over the world.

Economic development of many countries in gulf region is built on abundant availability of Petroleum products. The huge reserve of oil and petroleum products helped these nations to achieve financial dominance over the world. The growth index of these nations is strongly related to the production of petroleum products and then the motivating factor is to produce maximum profit out of it. If we proceed with the same utilization pattern of petroleum products, these sources will be depleted earlier than expected.

Gulf countries being the bulk suppliers of oil have gained financial stability and are in the top list of developed nations. The countries with good reserve of oil resources have become political powers. The enormous profit of petroleum industry, accrued wealth and political power eased up the unscrupulous exploitation of oil reserves. Even today these countries are dreaming of a bright future with their rich oil resources but these hopes are short -lived.

Easiness of conversion to other forms of energy made these sources very attractive that ended up in a global environment problem. Energy and environmental conditions are changing day by day. The entire world is worried about environment pollution and global warming.

After kyoto protocol the concern about environment is increasing among the international community. Now developed countries are forced to reduce power generation from fossil fuels and developing countries are struggling to reduce their dependence on coal and petroleum products.

Recent developments and scientific research has helped to reduce the cost of renewable energy and its conversion process. Among the available energy sources, wind and solar are considered to be most economical and easily accessible. The cost involved in setting up of wind farm projects is now drastically reduced.

More and more entrepreneurs are coming forward to reduce carbon deposits in environment and to promote renewable energy. The dependence on fossil fuels for energy purpose is considerably decreasing.

Each and every country is thinking of modification in their energy mix by adding more renewable energy. The reliance on renewable sources are sought after for fulfilling their energy needs. It is interesting to note the fact that all the countries in the world have great potential for either form of renewable energy.

But these sources are not being used or exploited to their fullest. If both the available wind and solar sources are effectively utilized, then they can alone contribute adequate supply to meet the energy requirement of entire world.

Normally the transition from one source of energy to another source takes decades. In the present situation, this transition period will be less for green energy. It may be noticed that switching over from fire wood to coal and then coal to petroleum products took over 50 years each. If the same length of time is to be considered for the transition from coal to wind and solar energy, then the time has come.

The acute demand of energy for human comfort and unscientific development plans has made environmental pollution and greenhouse gas emission to new heights.

Wind energy power plants will not cause any environment pollution and are more ecofriendly. Wind is replenished by natural processes on a sufficiently rapid time scale.

WAY BACK IN HISTORY

T he agrarian society settled by the bank of Nile river recognized the energy associated with wind. They effectively utilized this energy for the transportation purpose. In BC 3000 years Egyptians used wind sail ships to cross Nile river. They sailed wind powered rafts from one side of the river to other side with agricultural products.

Later in history people converted this kinetic energy of wind to mechanical energy and built windmills to grind wheat and other grains.

The wind technology slowly moved to Asia and Middle East. Arabian community started using wind power to grind grains. Apart from grinding grains and pumping water for irrigation wind mills excessively used for sawing wood, making paper and pressing oil out of seeds. The historical records reveal the significant development in agrarian age with the help of wind power.

Vintage model of simple windmills may have been

used in Persia (now Iran) as early as the 7th century AD. Iranians used these mills for irrigation and milling grain. The wheel bearing of earliest windmills were horizontally placed and supported by a vertical shaft. These machines were relatively inefficient.

These types of wind mills spread throughout the Middle East and were popularly used for agriculture purpose. Nevertheless, this type of windmill further reached to Europe. The earliest European windmills appeared in France and England in 12th century and quickly spread throughout Europe.

France introduced some excellent modification work on the crude model of wind mill and developed stone tower type wind mills. This stone tower consists of a wooden cap that supported the wind shaft. The rotatable wooden part of upper portion powered the windmill gearing and power transfer was achieved. This stone tower windmill construction took place somewhere in early 15th century.

Wind sails of an average length of twenty feet were attached to a horizontal shaft protruding from the end cap. Wooden planks were used as shutters to cover the sails and it looked like large paddle wheel. In post type wind mills the wind harnessing blades were rotated manually to orient the blades for accommodating maximum wind impact. Another invention was in fantail mechanism that could rotate the sails of wind mill orienting toward wind direction automatically. This helped a lot to enhance the efficiency of wind mills.

By the end of 18th century history recorded remarkable improvement in wind mill technology. The people of Netherland (Dutch) commissioned more than 10,000 wind mills during that period.

In the 19th century Denmark pioneered the technology for electricity generation from wind. This was a major breakthrough that introduced an alternative to challenge the basic dependence on fossil fuels for power generation. This laid the foundation stone for green energy revolution.

By the beginning of 20th century, United States of America widely employed wind energy projects for electricity generation. The rural belt of the western settlements was powered with electricity from Wind farms. The remote rural villages away from national grid were electrified with local grids powered by electricity from wind.

The rural communities were supplied with electricity from small wind turbines till 1940. Then the national grid was constructed across the country. Gradually technology was up graded and large wind turbine occupied the scene and in 1941 Smith-Putnam   generator installed at Grandpa's Knob, near Rutland, Vermont. The rest is history and United States took the lead in wind turbine technology.

Wind turbines operation is free from pollution, wind is renewable and it is freely abundant across the globe. This makes wind energy a global energy alternative to conventional fossil fuels. The remote villages and isolated hamlets of the world can be easily supplied with electricity from wind farm projects. Wind energy along with solar power is considered as a viable and valuable alternative for present world energy crisis in the face of global warming.

BACK TO BASICS

In our solar system, sun converts matter into energy every second. This converted energy reaches earth as solar radiation. Solar radiation contains both light energy and heat energy. The heat energy associated with sunlight falling on the earth induces temperature difference in atmospheric air and produces large scale movement of the atmosphere air.

In another words the differential heating of air above earth due to heat radiations from sun is the reason behind the formation of wind. The unevenly heated air moves from equatorial region to polar region. Thus the heat energy associated with solar radiation is converted into kinetic energy of moving air. This kinetic energy connected with moving air is wind energy. Thus wind energy can be defined as the kinetic energy associated with the movement of atmospheric air due to differential heating.

The axis of Earth is tilted at an angle of 23.5 degrees to

vertical and this decide which part of the world get more sunlight. The intensity sunlight at higher altitude and seasonal variation of atmospheric heating decide wind flow on Earth.

The self-rotation of earth on its own axis and inclination causes the wind to take direction around earth. Equatorial regions get heated up easily and air becomes lighter in weight and starts raising upwards. At the same time air in polar region, which is much heavier starts to fall downwards. The raising air at the equator moves northward and southward. Heating up of sea water due to sun light will affect the wind direction to a small extent.

The movement of wind is also affected by the type of the terrain. Mountains, valleys, obstacles in the surroundings such as buildings and trees, also has an important role on direction of wind flow. The power of the wind is proportional to the cube of the wind speed. It is therefore essential to have detailed information of the wind velocity and other characteristics associated with wind to calculate wind power.

The performance of wind turbines can be estimated accurately based on these wind flow data. The various parameters we need to know about the wind flow are mean wind speed, directional data of wind, wind velocity variations and height at which wind flows above the sea level. These parameters are used to assess the performance and economics of the wind power plant.

The annual wind speed at a location is a useful information as an initial indicator of the wind resource. The relationships between annual mean wind speed and wind potential status is given below

Wind velocity	Wind potential
0 to 6 m/s	Not Good
6m/s to 7 m/s	Average
7m/s to 8 m/s	Good
8m/s and 15m/s	Very Good
0 to 6 m/s	Not Good

There may be some locations for which wind data are not readily available. For such locations a qualitative indication of annual mean wind speed can be inferred from geographical details of the location by visiting the site.

The density of air is also an important factor in calculating the wind power. The density of air varies from place to place and again from season to season. Density of air is related to the local geography of the place and climatic conditions.

The velocity of air varies with altitude due to surface aerodynamic drag and the viscosity of the air. Due to rotation of earth the wind speed reduces at sea level but increases at higher altitude.

The wind speed increases with height and rises proportionally to the seventh root of altitude. So it is inferred that if we keep our wind mill hub at high altitude maximum power from wind can be extracted. Under cold climatic condition the air is still and wind speed is low. The stable atmosphere is disturbed when wind gains momentum at high altitude.

The selection of location for wind turbine should be with utmost care. Local wind condition is strongly affected by tall buildings, industries and geographical layout of that area. Wind shadowing effect, tunnel forming effect and surrounding vegetation also play an important role in setting up of a wind mill. The selection of site for wind turbine should be done by avoiding shadowing effects and turbulence effects.

Wind induced deformation of vegetation is true indication of wind potential for given region. Soil erosion pattern and rounding of stones or rocks give some valuable signals about the terrain which help in decision making regarding the location for wind mill projects.

More scientific studies can be conducted once the predictive information is positive in the proposed location. For accurate data wind speed measurement for two or three years is required. An anemometer data is a reliable source for further analysis.

The investigation for wind farm starts with location survey. The type, nature and distribution in terms of their soil condition, morphological, electrical and civil engineering aspects of proposed location are to be studied. The wind farm location is classified according to availability of wind as zone 1, zone 2, and zone 3 considering the suitability for wind turbines. The availability of land, wind speed, wind density and transport facility to the location decide the feasibility for wind projects.

Other important parameters for windfarm location study are slope of the land, water clogging on surface, drainage facility and other criteria such as proximity and facility for power evacuation. Steep surfaces are not suitable for wind mill erection. The slope should not exceed more than 20%. Secondly the earth resistivity and soil strength are also to be measured.

Generally, most promising wind locations are seen at hill zones gaps and offshore. The wind gaps are formed due to time long deformation along the major hilly regional zone. If the gaps are oriented perpendicular to the wind direction, they form potential location for trapping the wind power.

A wind energy generator installed at potential location cannot absorb all energy of wind that is flowing through it. It is subjected to some limiting factors and turbine design parameters. A German physicist named Betz conducted serious research on the conversion efficiency and calculated the possible hypothetical value of energy that can be generated from wind.

The maximum possible kinetic energy abstraction from wind is 59.3% of total kinetic energy. This constant is defined as coefficient of performance (C_p) of wind turbine. Betz's law specifies the energy extraction from open flow of wind through the rotor blades. The design parameters of wind turbine have no relation on this limiting factor. No turbine of any given technology can extract more than 16/27 (59.3%) of the total energy associated with wind flow. Betz's coefficient is calculated to be 0.593 applicable to all modern wind machines.

ENERGY HARVEST: WIND FARM

W ind farm development activities start once the location search is completed. A number of wind turbines are installed at particular area for large scale production of electricity. These types of arranged wind cluster are usually called a wind farm. The arrangement of wind turbine is based on certain scientific configuration according to wind flow pattern.

The mapping procedure is done by analyzing the information of geographical data and wind potential of that area. This data is available on websites published by different international agencies. Next step is field validation of wind farm areas mapped with the help of computer software.

The mapping procedure is done by analyzing the information of geographical data and wind potential of that area. This data is available on web

Construction of Pooling station for Power evacuation

The mapping procedure is done by analyzing the information of geographical data and wind potential of that area. This data is available on websites published by different international agencies. Next step is field validation of wind farm areas mapped with the help of computer software.

Once the computer modeling is done, then the data is to be validated for accuracy. Before establishing wind farm at a given location the results of wind energy mapping study in combination with remote sensing data is subjected to field verification.

In a wind farm large number of wind machines are commissioned. This is a tough job as the wind shadow of one machine should not fall on another. The machines are arranged in such way that each machine should get sufficient wind to generate its maximum capacity of power. Any given machine should not cause any interference to other machines in its neighborhood.

Micro sitting of wind turbines

The science of limiting array effects on wind turbine by a suitable geometric arrangement of machine layout is generally known as micro sitting of wind turbines.

Deployment of wind turbines in limited boundaries causes extra turbulence in close proximity. This sudden change in wind environment causes reduction of power generation. On some occasions, the wind speed will fall along with wind instability.

Air Turbulence effect reduces the service life of wind machines. Increased turbulence will damage the blades of wind turbine. The micro sitting exercise should be carried out keeping this extra caution element in mind. When wind mill is exposed to cyclic stress the humidity damage that may cause to various components decides the critical life factor. There will be fall in power generation and small drop in voltage output of machine due to wind turbulence.

In a well-designed wind farm no machine suffers drop in power generation due to the interference in wind flow caused by other machines in a given array. If all the machines are under single ownership, the group efficiency of the entire farm as one figure can be taken. But if different individuals or different companies are together setting up machines in a wind farm, then it will be very difficult to assess the efficiency of individual machine.

The cycling effect of wind turbulence induces machine shadowing and that will cause drop in power output of wind generator. Multiple ownership at a given wind farm area will invite dispute among owners due to the turbulence effect. Hence windmill micro sitting and array arrangement in difficult and sloppy terrain require multidimensional mapping.

Optimization of available land for installation of maximum wind power generation considering the geometry condition and terrain condition will determine layout pattern. The major elements that establish the efficiency of any wind farm is as follows

➤ **Wind velocity and predominant wind directions of given location**

➤ **The proposed machine design param eters**

➤ **The slope of the land**

➤ **Number of machines that can be accommodated in a particular place**

➤The power generation capacity of　　proposed machines

➤ The rules and regulations of local authorities and other land restriction factors of given locality

Once the above technical parameters are available wind farm optimization activity starts. The maximum power output from that wind farm can be calculated with above data with the help of computer software. The technical specification and details of wind turbine should be made available by the wind turbine manufacturer. The spacing between adjacent wind machines should be nine times of its rotor diameter along wind flow direction and five times perpendicular to the wind stream track.

As height of wind turbine increases, the accessibility of high velocity airstream increases. Air stream strike on high altitude turbine blades is more and maximum power generation is possible. Selection of rotor system with large blades and tall tower structure will result in high power generation. Final decision on wind farming can be taken after optimizing the cost of installation and yield.

Selection of wind turbines

The commercial viability of wind farm can be assessed

with the help of wind resources data published by international wind research agencies. Based on this data selection specification for wind turbine can be finalized for financial and commercial viability.

The general operating temperature range of wind turbine is from 20 degrees Celsius to 40 degrees Celsius. Special type of machinery is required for extreme weather condition. If atmospheric temperature is below 20 degrees Celsius, special construction materials are required for rotor blades. In such cases turbines should be well protected from ice falling. The international standard accepted for wind turbine construction is   IEC 61400.

At low temperature anemometer reading may become inaccurate and control system of wind turbine may become ineffective. Recently, wind turbine manufacturers offer low temperature machine with internal heating arrangements.

In such machines special type of lubricants are required for cold weather condition. The construction materials of these turbines should be of special alloys.

In winter, if the wind is intermittent or of low speed, then wind turbine require external power supply to heat up the internal parts.

Site preparation

The first step in site preparation of wind farming is contour map drawing. The presence of objects like tall trees, wind turbines by adjoining developers, power transmission towers or high rise buildings should be marked in contour map.

Site preparation for wind farm

The sites proposed shall be assessed based on the methodology recommended by the relevant international standard. If the slope of the land is more than 10 %, the type test cannot be conducted for machine. The geographical conditions of terrain, environmental and electrical grid accessibility should be favorable in order to carry out the above type tests as per the recommendations of the IEC standards.

The sites proposed shall be assessed based on the methodology recommended by the relevant international standard. If the slope of the land is more than 10 %, the type test cannot be conducted for machine. The geographical conditions of terrain, environmental and electrical grid accessibility should be favorable in order to carry out the above type tests as per the recommendations of the IEC standards.

The electricity grid connectivity should be tested for electrical safety and to safeguard from lightning. The electrical equipment should be type tested as per recommendation of IEC standards.

Meteorological mast

Positioning of meteorological mast require special study during the site visit. The slope of terrain gives an indication for prepositioning of the mast. During field visit site engineer will collect following data.

➢ **Identify the obstacles that may hinder the flow of airstream.**

➢ **Exact position for fixing the Meteorological mast.**

➢ **Social and environmental issues about the location should be collected.**

➢ **Permission for the installation of Meteorological mast.**

➢ **Accessibility of electrical grid connectivity.**

➢   The facility for testing of data collected by the Meteorological mast.

Security of the testing equipment from wild animals and from theft

Power evacuation: Wind mill transformer station

The evacuation of generated power from wind farm is another area of concern. As wind farm transforms the kinetic energy of the wind into electrical energy, it is to be transmitted to the load center. This requires construction of transmission lines and power pooling stations. The power from different pooling stations is transmitted to wind energy substations. All these require power evacuation feasibility study report before starting the projects.

Many potential windy sites are still not utilized due to lack of grid connectivity.

Construction of electricity transmission network from nearest substation to wind farm site involves cost and labor. The power generated from different machines of wind farm is pooled at a single point and from there transmitted to the substation through shortest feasible route.

Wind is available during monsoon months and solar is available mainly during summer months. Hence the same electricity network can be used for evacuation of power from both projects.

Solar panels can be positioned inter space between wind turbines or nearby vacant land with common power evacuation facility which

make project more feasible. The hybrid type of wind solar project also can be considered as wind and Solar are mutually complementary.

The wind farm technology was mysterious and very expensive at the beginning. But now it is in par with technologies of electricity from conventional fuels. It is worth to mention that many locations around world have untapped potential for wind farm. Some potential wind farm sites of the world are not professionally utilized to harness power due to lack of knowledge and technological guidance. New initiative for such projects not only solve the power crisis but also it caters employment opportunities.

WIND TURBINE: THE WAY IT WORKS

The modern wind turbine is a multifaceted, integrated system and its working is controlled automatically by microprocessors based control system. The major components of modern wind turbine are tower, nacelle and rotor. The tower supports the wind turbine and associated equipment. Rotor with blades helps to converts wind energy into mechanical and then to electrical energy. Nacelle accommodates the gear box mechanism and generator of wind turbine.

Wind turbines are of two types. Horizontal axis wind turbines and Vertical axis wind Turbines. Most common wind turbine is the horizontal axis (HAWT). In horizontal axis wind turbines, the rotor will spin parallel to the surface. In vertical axis wind turbines(VAWT), the rotor will spin perpendicular to the surface. Ninety percent of modern wind turbines are of horizontal axis wind turbines.

In horizontal axis wind turbines, the soft spinning rotor shaft absorbs kinetic energy from airstream and converts to mechanical energy. This mechanical energy is converted to electricity with the help of a generator. This is the basic working principle of a horizontal axis wind turbine.

In a cost effective wind turbine design model, all parts must be reasonably priced, lightweight and long lasting. The turbine should with stand all weather conditions and can be manufactured easily for uneven loading. Turbine systems that have less failure rates, require fewer repairs will lead to declining the cost of wind energy. The percentage cost split up of key parts of a contemporary turbine is given in table.

Tower	22% (percentage of total)
Blades	18% (percentage of total)
Gearbox	14% (percentage of total)
Generator	8% (percentage of total)

Tower

Tower of a wind turbine is the supporting structural equipment. These Structural elements embrace the best part of the weight and cost of a wind turbine. Tower and other supporting mechanism of wind mill should be strong enough to carry the entire weight as well as the dynamics of wind.

The power cables and control cables coming out of nacelle are brought down to ground along the tower housing. Towers are of two types, lattice towers and tubular towers. Large and medium size wind turbines are constructed with tubular structure, while lattice structure are more economical for small wing turbines. Tubular towers are preferred for large turbines of easily accessible location. Lattice tower is often used for small wind power turbines where transport facility is limited.

The tower can be constructed either with steel or with reinforced plastic materials. As the tower height increases the diameter of tower base should be large enough to accommodate more loads and to avoid the buckling effect. Towers are further classified to offshore towers and land towers.

Structural design of wind turbine foundation is very intricate. It requires considerable attention as it has to bear the static and dynamic load of wind turbine in operation.

The picture shows the civil foundation work of a tubular Tower construction at Kanjhikode site in India. The maximum feasible angular deflection of foundations during wind mill operation should be less than 0.5 degree.

Rotor

Rotor is the device which absorbs kinetic energy from the winds. The absorption of kinetic energy is proportional to the swept area of blades. Large sized rotor blades can absorb more kinetic energy from wind. Large sized rotor blades lead to cost escalation of construction and maintenance of wind turbine. This will further invite additional cost investment in tower foundation and tower structure work. At the same time rotor blade should be lightweight and strong enough to withstand against the speedy winds.

Economical wind turbine design suggests a rotor diameter of hundred meters or less. Mega wind turbine of recent installation has rotor diameter of 175 meter or more

Number of blades decides the efficiency of a wind turbine. More number of blades helps in absorbing more kinetic energy from wind but the cost of construction also increases. Considering the manufacturing cost of multi blade wind turbine and investment return from power production, number of blades ilimited to three for optimum efficiency and cost of construction.

The wind turbine blade count also determines system reliability, basically the dynamic loading of rotor. Wind turbine rotor blades are always aligned towards air stream with the help of yawing mechanism. While aligning the wind turbine, each blade will experience asymmetrical stress because of its positional difference. Three blade systems are more balanced than other blade designs.

The wind turbine blade count also determines system reliability, basically the dynamic loading of rotor. Wind turbine rotor blades are always aligned towards air stream with the help of yawing mechanism. While aligning the wind turbine, each blade will experience asymmetrical stress because of its positional difference. Three blade systems are more balanced than other blade designs.

Blade materials

Material selection is another vital factor in rotor blade design. In olden days, the wind turbines blade material of heavy metal such as steel were used. This causes high inertia to the rotating system but the advantage is that fluctuation in wind speed could be offset with high inertia. Another material used for wind turbine was canvas material. Using canvas material for construction of windmill blades limits the aerodynamic design. The attraction for using cotton canvas material is due to its low cost and wide availability.

Flat design of blades with wood or canvas material will cause low aerodynamic efficiency and relativity high drag to force ratio in capturing energy.

Aluminum or any other light material are always used which offer low rotational inertia. Light materials will always reduce the starting stress on rotor that can adapt to intermittent speed variations quickly.

The noise produced from a windmill is linked to the aerodynamic design of windmill blade. There is a strong relation between wind mill blade design and noise produced especially during abrupt stalling.

The manufacturing materials of turbine blades should be of low cost and easily available. Strong wind with stand capacity is an essential criterion for selection of wind mill blades. The ability to overcome lightning strikes and temperature rise under humid conditions are other factors that decide the selection of blades.

In modern wind turbine fiber blade are used. Fiber

blades are lightweight and strong enough to with stand all weather conditions. Working of a wind turbine at optimal slip ratio is more important in energy conversion during vigorous gust of wind. High inertia of rotating parts always buffers the slips in rotational speed and that will offset the changing turbulence effect.

Nacelle of Wind Turbine

In most of the Wind turbine the rotor rotates at constant speed matching to the frequency of electricity grid. But in most modern wind turbines, technology permits the rotor to rotate at different speeds. The power generated is synchronized to electrical grid with help of advanced techniques.

Nacelle with Generator

In large wind turbines the generators are accommodated in the Nacelle of wind turbine. The rotor of the generator is connected to wind turbine blades and that is allowed to rotate inside a stator which has field windings. Electricity generated at the stator is connected to the electricity grid with help of connection cable.

Generally, wind turbines are equipped with induction generators. An induction generator will consume reactive power from the system thus causing inefficiency in the system. To achieve unity power factor wind turbines are best equipped with capacitor bank at the bottom of wind turbine tower.

Wind Generator: Nallasinga

An electrical system is reliable and efficient if the wind

power generated is at unity factor. To achieve unity power factor a permanent magnet synchronous generator can be used in wind turbine. It has become a common practice in modern wind farms to use variable speed wind machines.

Whenever the wind speed goes below the rated average speed the torque generated from generator will help the rotor to attain the required speed. Field control of the synchronous generator is another method to regulate the speed. It works as a closed loop control system with a feedback.

Gearbox mechanism

The gear mechanism along with its control system is assembled in nacelle of the windmill. With the help of gearing mechanism speed can be increased to higher speed. Practically the rotating speed of windmill blades are very slow. Wind turbine generator requires higher speed than windspeed to produce electricity to match with system frequency.

Usually speed of wind turbine blades are about 20 RPM. But speed of generator should be 750 RPM or more. Hence the gear box mechanism is introduced in between rotor hub and generator.

Pitching mechanism

The operation of wind turbines is intended for a range of wind speed starting from 4 meter per second to 25 meter per second. If the wind speed exceeds the limit then the machines has to pitch out the blades and come to stand still. This is achieved with the help of a pitching mechanism. The pitch control will precisely align the blade for maximum airstream impact.

In modern machines electrical servo motor are used to furl the rotor blades. Individual pitch motors are used for increased energy capture which is achieved with the help of full span blade pitch control.

Yawing mechanism

Yawing control equipment are accommodated inside

the nacelle. Yawing mechanism is required to keep the rotor pointing into the wind direction. Wind vanes are generally used in wind turbines to judge the direction of wind and are installed at the back of nacelle. The power output can be maximized with the help of yawing mechanism by minimizing yaw angle.

Breaking Mechanism

During maintenance of wind turbine and at the time of power restriction wind mill has to stall. The braking of machine is done by reducing the angle of attack of air stream on blades. Electrical breaking and mechanical breaking system are the two types of breaking methods widely used in wind turbines.

WIND ENERGY: BEYOND SEAS

T he wind energy was expensive at the time of its beginning. But now it is competitive with electricity from fossil fuels. It is worth to mention that many countries around world have changed their focus to wind farm projects to mitigate environment pollution.

Wind farm projects not only substitutes the environment solutions but it caters electricity to the remote village were reach of grid power is far from reality. The cost of wind turbine is also decreasing and which makes wind energy more affordable and financially viable. In developing countries, wind energy systems offer unprecedented opportunities to accelerate the transition to modern energy services in remote and rural areas.

In particular, with the rapid decline in the price of wind turbine technologies in the past decade, distributed power generation with wind sources are quickly emerging as the most cost effective. Wind has become a reliable and affordable energy source to provide modern energy to the rural village dwellers who are deprived of electricity connection.

Wind energy generated in rural areas will be consumed then and there itself. This will reduce the transmission, distribution and cost of electricity distribution utility. The added advantage is the reduction in transmission and distribution loss of electricity. Power generation at rural area will ensure increased access to affordable lighting, sustainable cooking, heating devices, communications, and refrigeration for rural households.

Last decade was a thriving period for wind industry. The power generated from wind farms in the United States alone is more than enough to meet the demands of electricity in a city as large as Chicago. The power generation from wind farm all over the world has multiplied many folds during last decade. As per world wind energy statistics the installed capacity up to December 2019 is 591GW which is a good figure compared to total power consumption of the world. The cumulative capacity of wind turbine installation in USA alone is more than 100 GW. The rise in capacity addition for the last decade was 400%.

At present the global capacity of wind generation is more than 560 GW. China, United States of America, Germany, India, Spain and United Kingdom are the major countries that generate wind power in the world. Technology advancements and a dramatic decrease in costs have made wind energy broadly cost competitive with conventional fossil energy sources.

China leads in the chart of top wind power producers of world with a 240 GW wind farm capacity followed by United States of America and Germany. China has made tremendous improvements in wind energy during the last decade. Jiuquan wind farm of China is the biggest wind farm of the world with the installed capacity of 20 Gigawatt. More than 7000 numbers of different size wind turbines are installed at this wind farm.

Power evacuation from this wind farm is achieved with 750 KV High Voltage Direct Current (HVDC) transmission line. This transmission network is connected to the state grid of China and the entire electricity generated from this wind farm is transmitted to other parts of the country. This wind farm is of hybrid wind-solar model having solar power plants in between wind turbines. The total power generated from this plant is transported to eastern parts of China.

United States of America is the second largest wind farm developer, known for offshore wind farms. This country is having six of top ten offshore wind farms. In terms of installed capacity, United States is second to China with 96.4 Gigawatt of installed capacity. This is an offshore wind farm owned and operated by Terra gen power.

Alta wind energy center in California is of installed capacity 1548 megawatts and operational capacity of 1020 megawatts is one among them. Shepherds flat wind farm is the second largest wind farm in United States of America. The capacity of this wind farm is 845 megawatts and it is located near Arlington in East and Oregon. This started operation in 2012 with a cluster of 338 machines of 2.5 MW each. The electricity generated from this wind farm is supplied to California and the power generated from this wind farm is sufficient enough to serve the energy needs of 2 lakh house households.

Roscoe wind farm is located in Texas and is the third largest wind farm in the world. This 800 megawatt wind farm is developed around 400 square kilometers of farm land. There is 630 wind turbine of different generation capacity and the wind farm started its operation in 2009.

Germany being the third leading country in wind industry is having an aggregate installed capacity of 59.3 Gigawatt. Germany is number one wind energy producer in Europe and Gode wind farm of Germany has a combined capacity of 582 Megawatts.

After China, India is the largest wind energy producer in Asia. The total installed capacity of wind energy in India is 35 Gigawatts. Jaisalmer Wind farm of India is having 1600 megawatts capacity with wind mills of different sizes. Jaisalmer Wind farm is the biggest wind farm in India. This project was developed by Suzlon Energy Limited in Rajasthan. Machines of this wind farm is owned by independent power producers, public companies and individual investors but is operated by Suzlon energy Ltd. Different types of wind turbines have been installed at this station as according to the site condition.

Muppandal wind farm in Tamilnadu is another main wind project. Muppandal wind farm is having a capacity of 2500 megawatt and it is an onshore wind farm built in Tamil Nadu. This wind farm was developed by utilizing barren lands of Kanyakumari district which was not fit for cultivation or any other agricultural purposes.

This area known for high speed wind for continuous ten months in a year due to the presence of gaps in western Ghats mountain range. Apart from this a large number of small scale wind power projects are under construction in

India.

Spain has 23 Gigawatt wind energy installations. For the last few years there has not been much improvement for capacity addition in wind energy sector due to some policy decisions.

Main wind energy source of United Kingdom is from offshore wind projects. The largest offshore wind projects in UK is situated at the coast of Cumbriya, north west England.

Brazil, Canada and Italy are the other countries that generate wind power for the world. France is now moving from nuclear power plants to wind energy sector. The wind power installed in France has an aggregate capacity of 15.3 Gigawatt as on date.

In South America, Brazil is the largest wind energy producer with 14.5 Gigawatt wind projects in its credit. Market share of Canada is 12.8 Gigawatt. Italy is mainly focusing on wind energy fields in the south and on its islands. The present wind farm generation capacity of Italy is only 10 Gigawatt.

Thousands of engineers and technicians are working hard in inventing new methods and technologies to improve the efficiency of wind energy conversion. World is giving very much importance to wind energy development program. Every year, 15th June is celebrated as "Global Wind

Day". The coordination work for Global Wind Day is done by "The European Wind Energy Association (EWEA)" and the "Global Wind Energy Council (GWEC)". Network of partners of wind industry makes all arrangement throughout the world.

This day is dedicated to the pioneers of wind energy projects and to appreciate the efforts taken by our ancestors in promoting wind energy. Various activities are conducted on that day. Global wind energy day seminars and workshops are conducted to share the knowledge about wind energy among enthusiastic people during the meeting.

Students are allowed to interact with experts of Wind Technology and clear their doubts. The possibility of converting this world into a green energy world is explored.

WIND POWER: INVESTMENT OPTION

I n the present scenario wind industry is jammed with the COVID-19 pandemic crunch. Many countries postponed auctions and put off their timetable to start the installation work. Looking ahead, the COVID-19 crisis is to continue for few more months prolonging the targets for auctioning and erection results no project deadlines will meet in near future. The governments have to protect present and agreed wind projects by withdrawing from backdated changes against remuneration schemes.

Policy changes and guideline modifications are the need of the hour to revamp wind industry from total collapse. This is required to control the fossil fuel emission of greenhouse gas which has reached such a level that our Earth cannot accommodate any more.

Putting in place such revamped policy changes will be necessary to drive forward wind power development of the world. This policy stimulation can help wind industry to play a major role in creating jobs and investment in critical infrastructure. More promotional schemes are required to get economies back on their feet in post-crisis and make our economies and energy systems more resilient. This may help the wind industry to maintain its progress during yesteryears.

The year 2019 noticed substantial cost reductions in the wind farm machinery due to large scale production. The growing demand for wind turbines across the world has the made the manufacturing units more economical.

The growing recognition of new technologies and their cost effectiveness attract new players to the field. Apart from these the wind farm is playing a large and growing role in providing productive energy services and employment. The manpower is cheaper in rural areas and job opportunities has increased after wind farm projects.

The size of wind turbines keeps increasing by every year and the capacity has reached up to 10MW. Vestas has manufactured wind turbine of capacity 8MW in 2014 and now they have upgraded their technology to manufacture higher capacity wind turbines. In the same tune the average plant load factor has increased from 30% to 45% within one decade.

Continued economic slowdown in Europe and the COVID-19 epidemics in US made it difficult to make projections for 2021, which was just over previous predictions, not anticipating the dramatic growth in the Chinese market.

Market growth of wind industry is not up to the expectation in 2020 compared to 2019 due to covid-19. Only China could achieve the preplanned target of 300GW in 2020.The policy uncertainty in North America other developing countries caused a downturn for wind industry for past few years. Now the situation has changed.

Major players of wind turbine manufacturing industry are given in the table.

Rank	Company	Country	Total GW
1	Vestas	Denmark	9.60
2	Siemens	Spain	8.79
3	Gold wind	China	8.25
4	GE	U.S.	7.37
5	Envision	China	5.78
6	Ming Yang	China	4.50

The slow down process in Asia resulted due to covid-19 and is expected to be short lived. In coming years, global dominance by Asian countries in wind energy is expected. World leaders of turbine manufacturers are anticipating a bright future in 2021

During last centaury the cost of wind projects was roughly estimated to be $3000/KW and it fallen down to $1500/KW during 2019-2020. The post COVID -19 costs of wind projects will be still low.

Now United Nations Environment Protection program (UNEP) is promoting wind energy as there is no pollution issues. In wind farm project no fossil fuel is burnt to produce electricity. Hence this type projects can claim carbon credit. This is an attracting factor for investors if they have renewable energy purchase obligation (RPO).

By the end of 2020 world will have an installed capacity more than 650Gigawatt. It was expected a major breakthrough in 2020 but it has been dampened by covid-19 pan endemic.

The current corona virus crisis had a global impact on the market development in 2020, so that the wind industry worldwide will experience a general slowdown. Disrupted international supply chains and national lockdown regulations are both hampering the wind sector, like most other industries. The transition from conventional energy to renewable energy will not increase the financial burden on any country. At the same time, it will stimulate the socio economic growth and ecological balancing.

This is the right time for the countries to start development plans and stimulus programs to restore their economy after the corona crisis. World wind energy Association is advising member countries to invest in renewable energy and to orient economic growth strategies for a better renewable energy mix. It is really an admirable fact that more than 90 countries all over the world engage in power production from wind energy. It will be helpful to overcome the damage caused by Corona crisis. As an additional benefit, it will address the second challenge of climate change and thus wind energy remain in top position among any other investment options.

Technological innovation and low cost manufacturing of turbine equipment encourage entrepreneurs to this industry. Many countries declared tax benefits for the promotion of renewable power. Electricity utilities of many countries give special purchase deals for green power coming from wind farms, by attractive pricing mechanism. Hence this is the right time for entrepreneurs and developers to invest in wind energy sector.

WIND POWER: MYTHS AND FACTS

There are many apprehensions and fears in connection with the setting up of a wind turbine farm in a locality. People in rural areas especially view this with suspicion. Moreover, there are people in any locality who are generally disturbed by any change and try to find fault or spread unwanted rumors related to the setting of windmills. Nature lovers feel the beauty of nature might be spoiled and wind power generation is not natural.

It is the fact that this type of suspicion started long back in history of wind energy right from Egyptian period. There were objections against sailing ships and grinding grain by wind power. When Europeans developed wind turbines that could convert wind power to electrical energy, the subject of objection shifted to newer issues. Now sustaining all objections, wind energy projects have spread all over the world and wind energy became a major component in present world energy mix.

Wind Energy has the advantage of being harnessed in rural and remote villages, not being limited to big cities. Wind driven electric generators could be utilized as an independent power source, and for the purposes of augmenting the electricity supply from grids. In densely populated areas, decentralized production of electricity would help local industries, especially seasonal agro-processing industries.

Operationally, modern wind turbines are as reliable as conventional power plants. Most commercial wind turbines are also known for their longevity. Many turbines have been generating electricity since the early 1980s. Many American farm windmills have been in continuous use for generations, while some traditional European windmills have been working for last 300 years.

The pollution free, environment friendly and economical wind power may find a suitable place among green energy sources. In the present context, the most acceptable answer to the question of energy crisis in the face of global warming and environment pollution will be wind energy. The rising concern of petroleum and gas pollution will definitely pave the way for acceptance of wind power wide and far. It is considered as the most environment friendly energy source will not produce any kind of air or water pollution. Wind energy is a clean and renewable resource.

The environmental impact study on wind farm projects reveals that wind mill had some negative effect. In some places the glittering blades of windmills attracted flying birds into the rotor. It reported that some protected species of birds were trapped inside the blades and killed.

Another serious allegation was that the visual impact was deranged due to the eyesore arrangement of windmills in a beautiful landscape. The visual beauty of landscape can be maintained even after windmill erection giving due respect to the landscape by proper planning for retaining the scenic beauty.

Wind is present everywhere around the world. The speed and density of wind may vary from place to place. If the wind is having a speed of 10m/hour, then that location is suitable for wind mill installation. Wind energy is highly practical in places where the wind speed is 10 mph or above.

Wind is present everywhere around the world. The speed and density of wind may var from place to place. If the wind is having a speed of 10m/hour, then that location is suitable for windmill installation. Wind energy is highly practical in places where the wind speed is 10 mph or above.

Wind farm work in progress at Agali

Wind is present everywhere around the world. The speed and density of wind may vary from place to place. If the wind is having a speed of 10m/hour, then that location is suitable for wind mill installation. Wind energy is highly practical in places where the wind speed is 10 mph or above.

The setting up of wind farm projects in windy areas is limited due to spread of negative propaganda among public. This is mainly due to lack of awareness about the wind mill technology. There is lots of myth about wind energy among the public. These are causing huddles in building wind farms. Few of the wind farm myths are as follows.

➤ Most people don't like wind farms because it spoils the aesthetic beauty of nature. They believe the beauty of nature will be disturbed by installing wind mills.

➤ Wind turbines have deleterious effects on people's health. There is possibility of radiation coming out from wind mills causing health disorder.

➤ Wind turbines can kill animals, birds and marine life. There could be a limitation in natural flocking of migration birds along wind farm belt in the case of onshore wind farms. Swimming and breeding area of marine species will be disturbed due to offshore wind farms.

➤ Wind farms are identical, unpleasant and spoil the visual treat of land scape due to its mechanical manmade structures.

➤ Wind turbines are noisy and the humming sound from rotating blade is harmful to human ear.

➤ The popular misconception among public is that industries in cities will provide better job opportunities. Wind farm in rural areas will not provide enough jobs as wind mills require less maintenance. A lot of people will become unemployed if we change from conventional energy sources to renewable energy sources.

➤ Most advanced scientific investigation system available

on earth as on date cannot predict the wind flow pattern. Hence the dependability on wind power as a base load power source is limited. This factor often discourages wind energy for the functioning of continuous processing industries.

➢ Since wind energy data requires knowledge of the weather form and wind flow conditions on a long term basis, it may be difficult to collect the essential database. Therefore, in areas where a large amount of energy is needed, one cannot depend completely on wind.

In spite of all limitation, wind is the most promising and less polluting green energy source on earth. It is available in every continent, every country and every region.

Wind is an economical source of electrical power. Wind energy is becoming more and more popular as the cost of generating power from wind is coming down. With the rising price of conventional fossil fuel energy, renewable sources of energy are being looked into and wind energy is one such a green source of energy.

It is an acceptable fact that winds speeds vary by the time of day, season, and even from one year to the next year, hence wind energy is an intermittent resource. The intermittent wind condition and unpredictable wind flow pattern is a strong limitation of wind farm. The intermittent nature of wind flow and unpredictable seasonal change on wind flow dampen the use of wind power for primary energy source purpose.

Most of the wind farms around the world are connected to regional grids and that ensure grid stability and base load parameters of power plants. It often acts as secondary source of power for supplementing the energy demands. This problem of uncertainty can be tackled with a judicious mix of conventional power source and wind power sources.

However, the intermittent nature of wind energy does not affect consumers when wind turbines are tied up with large electrical networks or a national power grid. The large and stable national grid acts as infinite source and the dilemma arising due to interruption of intermittent wind energy can be solved.

Another area of concern is about plant load factor of wind farm projects. At windy sites, it is common for wind turbines to operate for nearly 8 to 10 months of the year. Even when operating, the wind may not be strong enough for wind turbines to generate at full capacity. There are many places which get biannual wind flow in equatorial region. These are very suitable economical sites for wind farm development.

Power plant based on fossil fuels has a plant load factor of 70 to 80 percent. But coal fired power plants which usually operate at an average of load factor of 75 percent of its capacity. In general wind turbines of different capacity designs work in a cost effective way on every windy location. Wind mills installed on windy sites operate at an average of 30 to 40 percent plant load factor.

During off season of wind flow, the dry wind power shortage is compensated by hydro power stations or from other conventional base load power plants. In the recent times, the solar power also helps in meeting the off-wind power demands. Such hybrid wind–solar system helps to provide reliable power supplies to the consumers. Remote villagers settled away from cities and those who rely on electricity from wind turbines often use batteries as backup system to overcome black out and brownout. This backup power sources provide electricity during off wind period.

Aggregate technical and commercial (AT&C) loss is minimum in wind farm projects. The distributed power generation concept is in practice for wind power. Generating power where it is consumed is often found a boon for wind projects. Instead of setting up large superpower station far away from load center, it is more acceptable to generate power in a distributed manner. This will reduce the cost of power transmission in addition to aggregate technical and commercial (AT&C) loss due to long line transmission.

A TALE NEVER TOLD

In this chapter I would like to share my own field understanding with my readers. I started working with the wind energy project soon after my graduation in electrical engineering. My hand on job involvement with wind farm construction ranges from old fashioned machines to modern machines. My experience with the wind farm projects in different regions was unique in nature.

The concerns of one wind farm location is absolutely different from another one. This is common in every state within the country and beyond borders.

I was closely associated with wind farm projects in my home town in Kerala, "God's own country", a southern state in India. I had the chance to study about the productive use of wind energy. I was captivated by its functioning and my curiosity drove me to observe and study the entire project activities from the beginning to its completion.

While doing so, I encountered some fundamental problems related to wind technology and my senior consultants helped me to solve. With great enthusiasm and curiosity, I wrote letters to various manufacturers of wind mill enquiring about the know-how of wind turbine operation and the availability of required information.

The response was surprising and I received plenty of material from international consultants and organizations, who were willing to offer all kind of assistance I needed. This helped me to address a lot of environmental concern from public while executing the project at Agali site.

Wind turbines create a lot of noise which indirectly contributes to noise pollution. It is true that wind turbines produce continuous swish sound and is a common phenomenon. But on many occasions, sub- standard engineering and installation procedures pose trouble. At one site very near to village households, we were developing a wind farm with multiple number of wind turbines.

As we progressed with our construction activities, the hissing sound coming from one wind turbine was unusually loud and that became an issue among the local dwellers.

After few days' trouble shooting, it was established at last that there was an alignment problem in the machine. We were forced by villagers to decommission that machine and to abandon that location. Thus the issue was solved but the looser was the investor.

When I was working at Agali wind farm site, a group of youngsters approached me and they expressed their fear of cancer due to high frequency radiations coming from wind turbine. Though I was aware that it was baseless and meaningless, I could not convince them that there was no scientific proof related to this myth.

As agreed upon, we had a public hearing function in which medical professionals attended and explained the functioning with scientific data and the issue was solved. It is very essential to convince the local public before we start the work.

Once I had a very hot argument with activists of human protection council and when they disrupted our construction activities at site. They produced a research

report by a pediatrician named Neena pierpond al-
legedly identified an illness named "wind turbine sleep
syndrome".

As soon as this report appeared in the newspapers, the
local villagers started reporting sleep problems. Government authorities sought clarification from our side about
this newspaper report.

Even though there was no proven scientific study report
relating to loss of sleep due to proximity of wind turbine,
we were forced to shift our wind turbine from that location. Again the looser was one who invested his hard earned
money for the project.

Few days later, an audiologist spread a rumor, which was
reported in television channels that low frequency sounds
coming from wind turbine while in operation caused health
problems among pregnant ladies. Later on all these doubts
and fears were cleared but the losers on all occasion was the
wind farm developer. The projects were delayed for long
period with cost overrun.

This created dissatisfaction and lack of interest among
the investors of windmills as they could not claim the tax
benefits for that financial year of their requirement.

Various environment related tales that erupted after
the installation of wind mills at Indian sites are still

spreading among local people. The most convincing one among them was that the migratory birds had changed their route after seeing the rotating wind turbine blades.

Even after wind farm development activities, migratory birds flocked in greater numbers than previous years in all the well-known bird sanctuaries. Thus that myth also died out as there was no evidence of such shifting of migratory birds anywhere in the world where such wind farm turbines were installed.

Wind farms in rural areas always attracted lot of objection due to illiteracy of the people. Always the vested interest group exploited this lack of literacy and spread degrading stories to gain cheap publicity or for political mileage.

One of the myths, that spread in India was very interesting. The pseudo environmentalists propagated the pollination theory. If wind mills were installed near any forest area, then the pollination of taller trees will be affected due to absorption of energy from wind.

This gained large publicity and was much talked among activists. But they lost steam when international experts ruled out the possibility of such a theory with proven scientific studies.

Stories about variations in forming of clouds, thunder

and lightning was the next one. The interesting argument was that lightning will be attracted by wind turbine structure and may cause decline in yield. Fall in agricultural yield was also propagated more than anything and it caught the attention of people from different walks of life and mainly from farmers.

This apprehension was mainly among farmers in high range and to their surprise high range locations recorded maximum yield after   the installaion of wind turbines during the subsequent years.

There was a common misconception among people that once the energy from wind is extracted, that wind is worth for nothing and that may cause skin diseases.

Another myth among the public was that electricity is produced by wind turbine from water not from wind. They were very familiar with the conventional method of hydel power generation from primary school syllabus.

During the installation of wind turbines at a hilly region, deep holes were drilled to accommodate earth wire conductivity.

This activity was misunderstood by people that water was been pumped from underground for generation of electricity by wind turbine. The depletion of ground water in that area was very common at that time due to excessive boring of tube wells. This news gained publicity and caused another hurdle against wind farm development in my career.

A group of activists gave propaganda that the clouds may get scattered due the moving blades of windmill, thus causing shortage of rainfall. But luckily, wind farm locations recorded good rainfall after wind turbine installation which was a blessing for us.

These were some of many concerns, that we had the opportunity to acknowledge. The story never ends, even the future investors may have to face many more objections in case the local population is not informed and educated in advance.

WIND POWER: TIPS FOR INVESTORS

Wind energy revolution is a very big opportunity of future and many investors ready to experiment with it. The wind industry has grown from its initial starting period to a full grown phase and now offers many opportunities to invest. Wind industry investment is not risky but it requires tolerance. It is a long-term investment option.

In this chapter, I want to share few ideas, how best to get into this field as an investor. The demand for electricity is on rise, hence good option to invest in wind energy.

➤ **Pollution free energy source, environment friendly, ideal for those who dream high and at same time wish to conserve our mother earth**

➤ **Cost of wind turbine equipment are reduced and hence ideal for investment in post covid-19 era.**

➤ Value addition for neighboring property, hence attractive for developers who hold large extend of barren land.

➤ Accelerated depreciation on investment is offered by all countries makes the investment more attractive. Easy and quick payback options available now.

➤ Payback period is less than seven years for wind projects of biannual wind locations.

➤ Tax liability is reduced by investing in wind projects for those who earn profit from other business.

➤ Limited stock of coal and petroleum products. Wind is nonperishable hence ideal for long term investment.

➤ Wind is environment friendly and replenishing source and will remain as the energy source of future generation to come.

➤ Never go after out dated technologies. They may be inexpensive but will end up in inefficient harnessing of your precious wind potential.

➤ Wind farm creates job opportunity for poor rural villagers. Those who are interested in charity, it is better option by empowering the poor rural people.

➤ Selection of site is important as it ensures good wind

speed and wind availability.

➤ Electricity grid proximity is very crucial element in site selection. This decides the power interruption index level. The transmission loss and cost of line construction is another criterion for site selection.

➤ Transporting wind turbine parts during installation and for future maintenance require road approach.

➤ Slop of the land should not be more than 20% and preferably slope less than 10% for easy installation and to avoid site development expenses.

➤ Wind speed should be more than 5m/s and ideally should be 15m/s.

➤ Wind density is proportional to power output from wind farm.

➤ Eco fragility of the land may invite un wanted environmental issues at the time of wind farm construction. Take care during site selection

➤ Proximity of other turbines may cause wind shadowing effect.

➤ Air turbulence at given location will result in damage of machine blades and low power generation.

➤ Deformation of vegetation gives the indication of wind availability for purchase of land.

➤ Blunted peaks, conical peaks, shape of stones and ridges speaks out the wind flow pattern and is a true indication of tunneling effect.

> **Optimum capacity design of wind turbine and tower structure design for cost effectiveness and easy transportation make the investment option more attractive and cost effective.**

Investing in wind energy, require basic understanding about the wind farm operation. Selection of windy site, finalisation turbine specification and land purchase often done by developers having much bigger exposure to this field. But there are few investors who look into minute details of the project to get exposure of the industry.

There aren't a large number of components in a wind farm. Wind energy manufacturers are readily available in supplying everything from tower structure to rotor blades matching to the wind location.

Beyond purchase of wind turbine at a new location, a great option for investing in wind industry is holding shares of established wind farms. Investors can do that through contract agreement.

ABOUT THE AUTHOR

Dr. P. Rajan

Dr P Rajan did his graduation in electrical engineering and took his doctorate in 2014. He turned to writing during his stint as the officer incharge for wind farm projects taken up by the Kerala state government. Born in 1964, he gerw up in the beautiful scenic state of Kerala also called God's own country. He is at present a consultant for energy audit and renewable energy projects. He has first hand expereince in setting up of wind and solar projects.He has published books on Solar and Wind energy topics .

www.ingramcontent.com/pod-product-compliance
Lightning Source LLC
Chambersburg PA
CBHW021457210526
45463CB00002B/809